MY FIRST SCIENCE BIOGRAPHY



# Sir Isaac Newton

by Christine Webster

**Go to www.av2books.com, and enter this book's unique code.**

**BOOK CODE**

**AVW27767**

**AV² by Weigl** brings you media enhanced books that support active learning.

AV² provides enriched content that supplements and complements this book. Weigl's AV² books strive to create inspired learning and engage young minds in a total learning experience.

## Your AV² Media Enhanced books come alive with...

**Audio**
Listen to sections of the book read aloud.

**Key Words**
Study vocabulary, and complete a matching word activity.

**Video**
Watch informative video clips.

**Quizzes**
Test your knowledge.

**Embedded Weblinks**
Gain additional information for research.

**Slideshow**
View images and captions, and prepare a presentation.

**Try This!**
Complete activities and hands-on experiments.

**... and much, much more!**

Published by AV² by Weigl
350 5th Avenue, 59th Floor
New York, NY 10118
Website: www.av2books.com



Library of Congress Control Number: 2019938595

ISBN 978-1-7911-0938-7 (hardcover)
ISBN 978-1-7911-0939-4 (softcover)
ISBN 978-1-7911-0940-0 (multi-user eBook)
ISBN 978-1-7911-0941-7 (single-user eBook)

Printed in Guangzhou, China
1 2 3 4 5 6 7 8 9 0 23 22 21 20 19

062019
311018

Project Coordinator: Heather Kissock
Art Director: Terry Paulhus

Photo Credits
Every reasonable effort has been made to trace ownership and to obtain permission to reprint copyright material. The publishers would be pleased to have any errors or omissions brought to their attention so that they may be corrected in subsequent printings.

Weigl acknowledges Getty, Alamy, Bridgeman Images, iStock, and Shutterstock as its primary image suppliers for this title.

# Sir Isaac Newton

# Who Is Sir Isaac Newton?

Sir Isaac Newton was a scientist who lived more than 350 years ago. He is best known for his studies on **forces** and how objects move. He also studied light and color.

Isaac's research answered many questions about how the world works. His ideas changed how people saw the world. Isaac's work was the start of modern **physics**.

"To myself I am only a child playing on the beach, while vast oceans of truth lie undiscovered before me."

Isaac is known as one of the greatest and most important scientists of all time.

# Early Days

Isaac was born in Woolsthorpe, England, in 1642. His family lived and worked on a farm. Isaac's father died before he was born. Isaac was raised by his grandmother. Growing up, Isaac spent much of his time alone.

Isaac went to his local school. He was an average student but loved building models. When he was 12 years old, Isaac was sent to a school in a nearby town. He lived with another family at this time. When Isaac was old enough, his uncle suggested he go to Trinity College to continue his schooling.

Today, people can visit Isaac's family home, Woolsthorpe Manor.

## Where Is Woolsthorpe?

Woolsthorpe is a very small village about 93 miles (150 kilometers) from London, England. Isaac's childhood home in Woolsthorpe is now part of the National Trust. This group helps to protect important buildings or areas.

Trinity College was founded in 1546. Each year, the college has about 900 students.

# Starting Out

Isaac began studying law at Trinity College. He worked hard. By his third year, he decided to study mathematics and physics instead.

It was at this time that many people in England started to get sick. Some were even dying. This was called the Great Plague.

The school closed for a while, and Isaac returned home. Here, his ideas on how things move began to grow. The school reopened in 1667. Two years later, Isaac was offered a job as a math **professor**.

Isaac was a religious man. He wrote **more papers** on **religion** than he did on **science**.

# Influences

Isaac was born at a time when many discoveries were being made in science. These discoveries changed the way people thought. They also changed how people saw the world. Scientists influenced one another in different ways.

Isaac was part of a group of scientists called The Royal Society. This group still exists today. It supports the growth of science.

## Aristotle

Aristotle was an ancient Greek thinker. He believed that heavy objects fell faster than those that were light. Isaac was very interested in this idea.

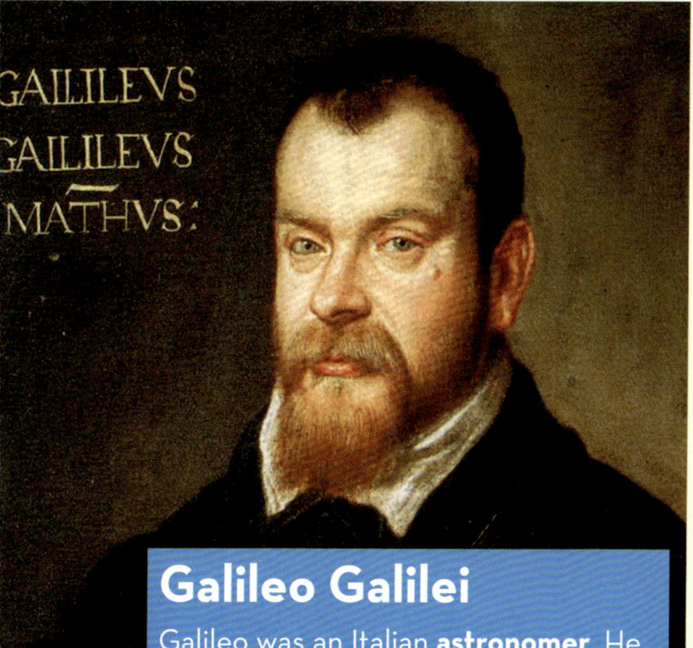


## Galileo Galilei

Galileo was an Italian **astronomer**. He believed that force was not needed for objects to keep moving. Isaac explored this idea further. He made it his first law of motion.

## René Descartes

René was a French scientist. Isaac studied his work at college. He had many questions about René's ideas.

# Practice Makes Perfect

In 1666, Isaac was sitting in his mother's garden. He watched as an apple fell from a tree and hit the ground. Isaac wondered what caused the apple to fall to the ground. He thought that some kind of force must have pulled it there.

Isaac began working to prove this idea. It took him 20 years to explain the force that made the apple fall. Today, this force is known as **gravity**.

Isaac published a book called ***Principia Mathematica*** in 1687. It took him **two years** to write.

Isaac was 23 years old when he watched the apple fall from the tree.

# What Is a Scientist?

A scientist is someone who studies things. People who study forces and movement are called physicists. Scientists are very curious. They love solving problems. Scientists try to answer questions through **experiments**. They collect information using a six-step method.

Isaac's experiments with light helped to explain how rainbows form.

# Using a 6-Step Scientific Method

STEP 1

## QUESTION

Scientists ask a question about what they want to learn. They read books and go online to research what other people know about the topic.

STEP 2

## HYPOTHESIZE

The scientists then guess what the answer to their question might be. This guess is called a hypothesis.

STEP 3

## EXPERIMENT

Scientists plan an experiment to see if their hypothesis is right. They gather the materials they need. Then, they set the materials up and do the experiment.

STEP 4

## OBSERVE & RECORD

Scientists use their senses to observe what happens during their experiment. They watch. They smell. They listen, touch, and even taste. They then record what they find.

STEP 5

## ANALYZE

Scientists think about what happened during their experiment. They decide if the experiment showed that the hypothesis was right or wrong.

STEP 6

## SHARE RESULTS

Scientists let other people know about their experiment and what they found out. They may write a report or give a speech.

It is said that while Isaac was trying to write a paper about gravity, his dog knocked over a candle, burning many of his papers. It took Isaac a year to recreate his work.

# Overcoming Obstacles

In 1672, Isaac wrote a paper about light and color. Most people enjoyed it, but some did not. Some scientists **criticized** Isaac's work. This upset Isaac greatly. He decided to stay away from other people. Starting in 1678, he lived alone for six years.

Isaac liked being alone. It allowed him to focus on his ideas. Isaac did some of his most important work during this time. He created his three laws of motion. They explain how things move.

Isaac invented a new kind of telescope. It used mirrors, instead of glass, to focus light.

# Achievements and Successes

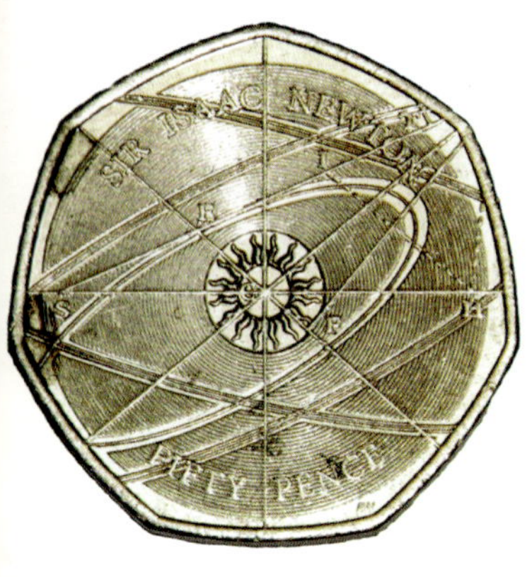


In 2017, the Royal Mint issued a coin in honor of Isaac and his work.

Isaac was only 26 years old when he became a professor. He worked at Trinity College for almost 30 years. Twenty years after becoming a professor, Isaac became a member of **Parliament**. This role let him help make laws for the country.

In 1696, Isaac became the **warden** of the Royal Mint. This is where England's money is made. Isaac worked there for more than 30 years.

When Isaac was 62, England's Queen Anne paid him a visit. She made him a **knight**. He became Sir Isaac Newton.

Isaac became president of The Royal Society in 1703. He held the position until his death in 1727.

Isaac's laws of motion helped scientists build the space shuttle correctly. The first space shuttle launch took place on April 12, 1981.

# Impact on Society

Isaac's ideas are still used by scientists today. His laws of motion play a key role in space travel. They let scientists know how much force is needed to push a rocket into space. His law of gravity helps them know how objects in space will behave when they are near each other.

Isaac also created a new kind of math. It is called calculus. Scientists use calculus today. It helps them explain force and motion.

Calculus is taught in high schools and universities around the world.

# Timeline

**1642**
Isaac is born on December 25.

**1661**
Isaac becomes a student at Trinity College.

**1687**
Isaac's first major work, *Principia Mathematica*, is published.

**1689**
Isaac is elected to Parliament.

**1727**
Isaac dies on March 31, in London, England.

**2019**
A **clone** of the apple tree in the Newtons' garden is planted at Clemson University, in South Carolina.

# Key Words

**astronomer:** a person who studies stars, planets, and other objects in outer space

**clone:** an object that is the exact duplicate of another object

**criticized:** expressed disapproval toward someone or something

**experiments:** tests that are done to prove a hypothesis

**forces:** actions that change the motion of objects

**gravity:** the force that makes objects move toward each other, for example, when things fall toward Earth

**knight:** a title that a king or queen gives someone as a reward for good work

**parliament:** a group of people who are responsible for making laws

**physics:** the study of energy, matter, and forces in the world

**professor:** a teacher, usually one of the highest-ranked at a college or university

**warden:** a person in charge of a building

# Index

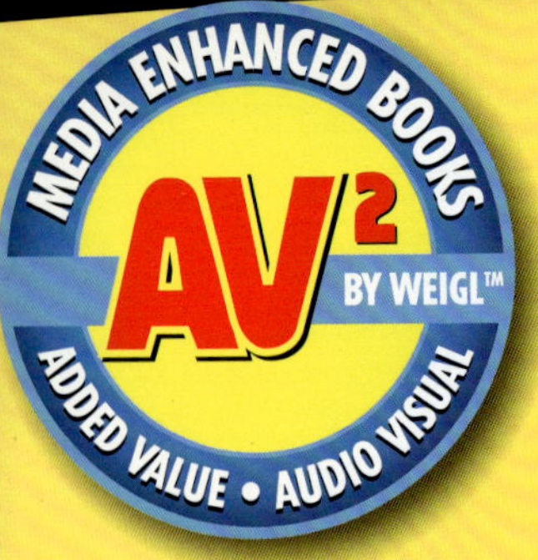


# Log on to www.av2books.com

AV² by Weigl brings you media enhanced books that support active learning. Go to www.av2books.com, and enter the special code found on page 2 of this book. You will gain access to enriched and enhanced content that supplements and complements this book. Content includes video, audio, weblinks, quizzes, a slideshow, and activities.

## AV² Online Navigation

**Book Pages**
AV² pages directly correspond to pages in the book.

**Audio**
Listen to sections of the book read aloud.

**Video**
Watch informative video clips.

**Embedded Weblinks**
Gain additional information for research.

**Key Words**
Study vocabulary, and complete a matching word activity.

**Try This!**
Complete activities and hands-on experiments.

**Quizzes**
Test your knowledge.

**Slideshow**
View images and captions, and prepare a presentation.

**AV² was built to bridge the gap between print and digital. We encourage you to tell us what you like and what you want to see in the future.**

**Sign up to be an AV² Ambassador at www.av2books.com/ambassador.**

Due to the dynamic nature of the internet, some of the URLs and activities provided as part of AV² by Weigl may have changed or ceased to exist. AV² by Weigl accepts no responsibility for any such changes. All media enhanced books are regularly monitored to update addresses and sites in a timely manner. Contact AV² by Weigl at 1-866-649-3445 or av2books@weigl.com with any questions, comments, or feedback.